NCA Report Series, Volume 3: Knowledge Management Workshop

NCA Report Series

The National Climate Assessment (NCA) Report Series summarizes regional, sectoral, and process-related workshops and discussions being held as part of the Third NCA process.

The workshop on strategies for knowledge management for the 2013 NCA was held in Reston, Virginia in September 2010. Volume 3 of the NCA Report Series summarizes the discussions and outcomes of this workshop. A list of planned and completed reports in the NCA Report Series can be found online at http://globalchange.gov/what-we-do/assessment.

CONTENTS

CONTENTS

U.S. Geological Survey National Center in Reston, Virginia
Photo courtesy of Joshua Davis Photography via Wikimedia Commons

Overview of the Workshop

The National Climate Assessment (NCA), under the auspices of the United States Global Change Research Program (USGCRP), convened a workshop on "Knowledge Management, Data, and Review Strategies for the National Climate Assessment" from September 20 to 22, 2010. The workshop was held at the United States Geological Survey (USGS) headquarters in Reston, Virginia. The purpose of this workshop was to begin to identify approaches and methodologies for managing the large quantities of data that will be either developed or redeployed in the context of the Assessment, as well as ways of archiving and retrieving that information. In addition, issues related to transparency, quality assurance and documentation were addressed. The overall goal is to ensure the quality of data used for the NCA, and efficiently manage these data in a way that makes them as usable and accessible as possible.

Attendance at the workshop was by invitation with a focus on government employees with experience in managing and deploying data (attendees are listed in Appendix B). Private and public sector representatives who had experience in addressing data issues associated with multiple previous national and international assessments were also invited to share their perspectives.

One of the desired outcomes from the workshop was to initiate the development of a community of people who have interest and expertise in managing information and developing approaches to data management to be considered by the Federal Advisory Committee (FAC) for the Assessment. People who participated in the workshop did so as individuals with no intent to provide any specific recommendations or consensus-based guidance.

This summary integrates the comments that were made by the invited presenters with the comments that arose in discussions associated with the presentations and in breakout sessions (see agenda, Appendix A).

Introductory Session

Major Challenges Identified

There are a number of key challenges that will need to be addressed in establishing the ongoing Assessment process as well as in writing the reports that the NCA will produce over time. These include

- How can we integrate and repackage data in a meaningful and comprehensive way?
- How do we ensure that the information that is produced is actionable and usable?
- How can we ensure that our decisions are well documented? There are very high expectations related to quality, and a significant amount of scrutiny should be expected.
- How do we ensure that both quality of data and relevance to decision makers are a priority? In other words, in our attempt to ensure that every fact is unassailable we should not ignore the fact that some sources of information may be less certain but actually more useful.
- How can we gauge how good the data are, and how mature? What criteria will we use to make these determinations?
- How do we compare and categorize data across different regions and sectors?
- What are our peer review responsibilities if we decide to include data that have not been previously published?
- How do we ensure public engagement? How do we respond effectively to public comments, especially if there are a large number of them?
- How will we be prepared for the fact that the relevant research questions and needs will change over time?
- What will it mean for organizations and agencies to partner with the Assessment, and how will we handle issues like potential biases, funding sources, *etc.*?

Major Themes

The issue of transparency was introduced at the beginning of the workshop and came up multiple times in reference to the process, the substance and data, the review, and documentation and response to public comments. The capacity to search review comments as well as the data themselves should be facilitated. The objective of transparency should always be foremost in our minds.

In a related but perhaps contradictory observation, it was noted that the burden on the scientists associated with peer review and documentation needs to be considered. This has become a major issue for participants in the Intergovernmental Panel on Climate Change (IPCC), since most of those who engage do so at a very large personal expense (measured in both time and effort).

An "end-to-end" approach is needed, in which the needs of stakeholders are identified and prioritized prior to designing the knowledge management system, and the utility of the information is evaluated over time in order to promote adaptive management of the data system. We need a sustained process with multiple products so that we are not starting the Assessment process with a new set of rules every four years. One way to ensure that the process is sustained is to have ongoing partnerships with non-government entities who also take some responsibility for monitoring and reporting information.

Other General Themes

- We have the opportunity to engage the public and help de-mystify climate change science; this is a high stakes activity that will require engagement of individuals with expertise in climate science communications.
- We need to keep collecting data and we must not lose support for long-term monitoring and observations, which form the basis for detecting change.
- We also need to take advantage of both new and existing monitoring efforts and indicators of change (*e.g.*, the USA National Phenology Network within USGS, the National Ecological Observing Network within NSF) and lessons learned from that work.
- We will have a much more significant challenge in data management with a live web-based data system as opposed to a one-time report (see full list of product in the NCA Strategic Plan at http://globalchange.gov/images/NCA/nationalassessmentdraftstrategy.pdf).

Short-term Actions Needed for Knowledge Management Activities

- Fine-tuning the overall scope and expectations of the NCA
- Development of a decision matrix of what data to accept
- Identifying key strategic partners both within and outside of the government
- Creating a strong internet presence and overall branding for the NCA

Other Assessment Processes

The workshop included presentations on previous assessment efforts, including U.S. National Assessments, reports on assessments from the National Research Council (NRC), the National Oceanic and Atmospheric Administration's (NOAA) annual State of the Climate Reports, and the recent InterAcademy Council Review evaluating the processes of the IPCC. The lessons learned from these previous assessments can provide input toward the developing NCA. Major knowledge management themes in previous assessments have been transparency and access to information.

Previous U.S. Assessments

Important input coming from previous National Assessment participants included

- It is not a wise use of resources to set up a separate archive for all of the information and data. However, all of the data and information must be traceable back to their original sources.
- It is not clear that the "build it and they will come" model of assessment work is useful (*e.g.*, in the First National Assessment, significant scenario and modeling data were developed but were not well utilized by the regional and sectoral assessment teams for a variety of reasons).
- The Second National Assessment effort, which resulted in *Global Climate Change Impacts in the United States* (2009), was primarily a synthesis of the literature with careful sourcing and referencing, with less focus on engagement or traceability of conclusions to the models or analyses from which they came.
- Regarding the process itself, it must be open and accessible and highly interactive. Individuals and organizations do not always have easy access to travel budgets to attend meetings, so there is a need to make use of remote access tools but also to remain sensitive to different abilities and needs to use them.
- Although there needs to be a presumption that legal challenges will occur, careful documentation and a transparent process will reduce the challenges.
- Regarding reviews of Assessment work, no anonymity should be expected. Reviewers should be identified and reviewer comments should be responded to (although it is unlikely that every comment can be responded to in a unique way). Establishing a search function so that comments can be scanned would be useful.

In summary, the most important lesson learned from previous National Assessments is the need for transparency, and the need to build a sustainable, ongoing process. An enormous amount of effort went into creating reports, but there were limited mechanisms available to update and share outside of the printed form. A further important lesson is that although one could manage all of the data within the NCA structure, it would be highly inefficient and prone to error to duplicate the existing data repositories. Much of the data used by the Assessment will have to be hosted by the people who generate the data, so a distributed knowledge management system will be needed.

IPCC Assessments: Lessons Learned and Recent InterAcademy Council Review

Controversies over errors in the IPCC Fourth Assessment Report (AR4) and the current debate and public scrutiny of climate science led to the InterAcademy Council review.[1] The committee made recommendations on both management and governance of the IPCC. Multiple specific recommendations were made about the process for decision-making over time (in between the plenary meetings of the parties) and the roles of the Executive Director, Secretariat and IPCC chair. Further recommendations were related to conflict of interest, and the need for stronger review and response to review processes to minimize errors and ease the burden on authors. In addition, it was recommended that a more consistent method for categorizing uncertainty be adopted, and a more sophisticated approach to communications be developed. Improved transparency in all stages of the process was a major overall recommendation.

NOAA's Annual State of the Climate Report: Lessons Learned

Because this report is issued annually, a challenge for NOAA's State of the Climate Report is managing the short time frame. A new report always needs to be in process shortly after the previous one is published. It was noted that non-meteorological data is not harvested with the same efficiency as meteorological data, and usually not on an annual time frame. This results in the need for NOAA to work with data providers on a case-by-case basis. NOAA is mandated to expand the breadth of the document by one variable per year, so they started by working with an oceanographer to infuse ocean

---

[1] http://reviewipcc.interacademycouncil.net/

and marine community data; with ecologists to include living systems and terrestrial systems, *etc*. Dealing with living systems is very different from dealing with data from the climate system, so it is more difficult to summarize status on an annual basis.

There is also tension between scope and utility. There are times when the relevance of the information is lost in too much detail. The use of reanalysis data in this annual assessment allows global analysis of poorly observed areas, but many consider this to be model input, which causes some issues of transparency and clarity. The reanalysis products are improving with each new generation of data.

An additional issue is navigating differing baselines for climatologies that come from different programs and missions. Which years should be used? The standards are different depending on where the data come from. Again, transparency is critical – reviews and comments and major data sets are placed online (*e.g.*, indicators).

Within the IPCC process, disagreements among authors have hindered progress. Refereeing disagreements and appeals requires a mediated process. One approach that might be useful for the Assessment is to designate an ombudsperson to act in conflict resolution situations. Also, it would be helpful to have "translation specialists" to help people from different disciplines understand each other. There is commonly a major language barrier between disciplines, but we are trying to generate interdisciplinary understanding.

The volume of data that will need to be archived is significant, so we need to create an archival template / metadata template for multiple steps of the process.

Additional insights from the discussions among the previous assessments panel included

- Assessment participants should not expect to reach a consensus on every topic. For example, the NRC does allow for ways to describe a range of views in their reports.
- There is a need to strike a balance between responsibly describing a range of scientific opinions and ensuring that errors or misstatements of fact do not get into the report.

We should be very clear about what we know, what is supported by the evidence, what is supported by models, *etc*.

- Describing how we know what we know – and traceability to the data – will allow us to show how we have come to specific conclusions.
- Our plan to do the Assessment in partnership with a group of people outside of the government requires a particularly clear "chain of custody" and documentation rules, as well as clear conflict of interest management.
- In the past, communications and education have started when the document is published; this approach is inadequate and needs to change for this NCA, with communications integrated from the beginning.

Information Quality Act and Highly Influential Scientific Assessments

A focus on the Information Quality Act (IQA) at the early stages will serve us well in terms of addressing data quality questions; these requirements are a common sense way of ensuring high quality data. The Office of Management and Budget's (OMB) guidelines are **minimum** expectations. The requirements include a pre-dissemination review. Agencies need to use their discretion concerning what the reviews should entail, noting that there are both costs and benefits associated with the reviews. NOAA/DOC has ultimate responsibility for the quality of information disseminated as part of the NCA. Each agency has its own guidelines which were reviewed by OMB and are consistent with the OMB government-wide guidelines. It is not a one size fits all approach. Key concepts are utility, objectivity, integrity, and reproducibility.

Not all information needs to necessarily be reviewed to the same level. Quality should be seen as a performance goal; the more important the information, the higher the quality standard that should be met. Normally, there is an expectation that data and conclusions are derived from peer-reviewed literature; peer reviewed literature is presumed to be objective, however the intended use of the information must always be considered. For *influential scientific information*, the data and methods must be sufficiently transparent such that the disseminated information (*i.e.*, analytic results) is reproducible by qualified third parties. The approach should be as transparent as possible and the data should be relevant and appropriate to the application. For example, a report that was developed and peer reviewed ten years ago for a different purpose may not be appropriate for current purposes. For peer-reviewed journal articles, there is a presumption of objectivity. The Peer Review Bulletin produced by OMB includes minimal standards for peer review: when, why, and what the agencies' responsibilities are for scientific information. The NCA is clearly in the *highly influential scientific assessments* category (a subset of *influential scientific information*), which entails ensuring that the highest review standards are met or exceeded.

Selection of the peer review mechanism depends on the type of information itself; peer reviewers must have expertise – then secondarily independence and absence of conflicts of interest. Balance amongst the peer review panel should also be considered. Reviewers should represent a diversity of scientific perspectives relevant to the subject; agencies need to ensure this. Reviewers cannot have taken part in the process, and they must produce a publicly available report of their findings. Agencies must have a peer review agenda, updated every six months. The agenda should list items that will be coming under peer review. There are specific requirements regarding the basic information about the review so that the public can comment on how the agency plans to review the document ahead of time.

In addition, for *highly influential scientific assessments*, to ensure independence, repeated use of the same reviewers should be avoided and agency personnel should not review documents from their own agency (*i.e.*, USGS should not review a USGS product even if it was produced by a different office). For these most important assessments, the agency should provide a written response to the peer reviewer comments explaining agreements and disagreements and any actions the agency will take in response to the peer review report.

IQA applies only to the federal government. The agency has to be responsible, and face the burden of ensuring that the data they are using meets IQA standards. Since NOAA is administering the Federal Advisory Committee for the National Climate Assessment, it has to meet the NOAA IQA guidelines at a **minimum**, but there may be additional requirements as an interagency product. Information that is included does not necessarily have to be in the peer-reviewed literature, although

agencies are encouraged to ensure that important underlying information has undergone appropriate peer review. There can be other peer review mechanisms. Outside experts may be able to do the peer review for this type of information.

Agencies have in place mechanisms to allow the public to request correction of information they believe to be non-compliant with OMB or agency guidelines. The burden is on the requestor to explain why they disagree with the information. Agencies must be responsive to correction requests.

Data and Source Material – Archiving and Access

The following presentations were included in this workshop because they are examples of large data archiving and integration efforts across the U.S. Government. There is considerable expertise in the staff of these projects and they are likely to be important resources for the Assessment. In addition, some components of these existing systems may be able to serve as distributed information sources for the Assessment data management system.

Program for Climate Model Diagnosis Intercomparison[2]

Data that comes out of PCMDI (Program for Climate Model Diagnosis and Intercomparison) feeds into the climate research community. Phase 5 of the Coupled Model Intercomparison Project (CMIP5) involves simulation output from 25 global climate

---

[2] PCMDI was established in 1989 at the Lawrence Livermore National Laboratory (LLNL), located in the San Francisco Bay area in California. It is funded primarily by the Regional and Global Climate Modeling (RGCM) Program and the Atmospheric System Research (ASR) Program of the Climate and Environmental Sciences Division of the U.S. Department of Energy's Office of Science, Biological and Environmental Research (BER) program.

The PCMDI mission is to develop improved methods and tools for the diagnosis and intercomparison of general circulation models (GCMs) that simulate the global climate. The need for innovative analysis of GCM climate simulations is apparent, as increasingly more complex models are developed, while the disagreements among these simulations and relative to climate observations remain significant and poorly understood. The nature and causes of these disagreements must be accounted for in a systematic fashion in order to confidently use GCMs for simulation of putative global climate change.

models. The data archive is open to everyone; there are 5,000 registered users and one petabyte (PB) of data available. Interest in the data continues to increase, with more downloads now than in previous years even though a new set of models is being developed for the next IPCC.

This intercomparison effort is collaborative across many governments and sectors including the National Aeronautics and Space Administration (NASA), the National Center for Atmospheric Research (NCAR), Lawrence Livermore National Laboratory (LLNL), and Oak Ridge National Laboratory (ORNL). The purpose is to build, operate, and support a global infrastructure for the management, access, and analysis of climate data. Data providers feed into four data centers, with portals (or gateways) at PCMDI, NCAR, ORNL, and NASA, (there are seven more coming soon). There are nodes (where data is stored and published) and sites (can be a node or gateway) in the system. OpenID allows users to search all of the gateways at once. If there are changes in the data after users have downloaded it, they are automatically notified.

With data versioning, users are able to access and cite any published object, as it existed at any point in the lifetime of the database. This allows for replication of findings. There are metrics that enable the data providers to trace who downloaded data, what data were downloaded, when it was downloaded, and from where, *etc.* There are provisions for transferring large amounts of data and documentation of provenance. There is an Ultra-scale visualization climate data analysis tool (UV-CDAT), an integrated framework that allows for analysis, visualization, comparisons, *etc.* These findings can be saved and given to users. The information from this sophisticated project is directly applicable to the Assessment knowledge management activities.

National Integrated Drought Information System

The National Integrated Drought Information System (NIDIS) program, though hosted by NOAA, brings data together from multiple federal agencies to help with water management in drought-prone watersheds. It provides a mechanism for integrating data into a decision-support platform, and is a good example of an "end-to-end" program because it has been developed in consultation with users and is constantly being evaluated and improved by people who depend on the information.

NIDIS started by focusing on several key watersheds including the Colorado (southwestern U.S.) and the Apalachicola-Chattahoochee-Kent (southeastern U.S.) basins where drought has been a serious concern. The lessons learned in these pilot studies are designed to be mapped to other regions. Part of the thought process for NIDIS and the Regional Integrated Science Assessment (RISA) teams is to identify existing threats in the context of today's climate variability, and climate-sensitive development pathways that might be put at risk (ecosystems, economies, *etc.*). As the climate changes, adaptation will need to happen at multiple time and space scales, and it may or may not be able to keep pace with the rate of changes. There are multiple tradeoffs that need to be made in the context of scale and certainty of climate information relative to the scales that decision makers would like to see.

There is a presumption on the part of data providers that if they provide information, it is useful – but how can we track performance of data and tools over time and know that a program is useful? NIDIS information is tested regularly by water managers and there is a feedback loop to ensure utility of the information for dam operations, drought management plans, *etc.* It is designed to support the drought "triggers" that states or others use to manage their drought programs.

Climate.gov

Climate.gov is the NOAA climate portal that is being designed as the "one-stop shop" for climate information across U.S. agencies.

There are two key questions in designing the interagency climate portal:
1) Who are our priority audiences, and what are their needs and wants?
2) What are our objectives for communicating with them?
It is intended that the climate.gov portal will be the primary way to engage communities in climate issues. How will we judge success? Do people know the site exists? Is the site trustworthy, usable, or relevant?

There is a two-pronged strategy for building relationships with the public related to climate. gov: dialogue sessions and the portal itself. There is also a phased approach to building the site; NOAA wants the portal to become a truly interagency effort and to include information from other agencies. It

will provide centralized access to data generated through a decentralized process.
Climate.gov is currently transitioning from a prototype phase to the operational phase and is a major priority for climate service activities across the U.S. government.

World Bank Climate Portal

The World Bank Climate Portal was designed to provide ready access to climate information, in cooperation with disaster risk management activities. It includes a simple expert system (ADAPT) to guide people to main threats (climate risks) in their region. Open source software was used throughout, which provides simple means of access to other data sources. The software basically uses Google tools with standard querying techniques. The portal also includes a socioeconomic database and can rapidly summarize, overlay, and visualize data. It links current trends in climate to climate projections and will soon expand to include a country adaptation profile and an operational tool, with sector-specific information. People are requesting "dashboards" that allow them to answer questions about climate change impacts in regions.

There are special issues in international projects – country-led constraints; stakeholder expectations that require significant negotiations; consultations with governments regarding their expectations (which may not be consistent with a scientific perspective); and complexities in using indigenous knowledge, which needs to be considered and included with care. As in other adaptation support applications, downscaling of global model output to local scales is a frequent request.

Rapid learning occurred within the World Bank as demand shifted to rapid information access for quality data. Users are becoming more sophisticated, but reviewing the knowledge can be problematic. Time and cost pressures, limited resources, and inflexible delivery schedules have interfered with progress.

The use of grey literature is also very complicated and remains a contentious issue. Judgments are made on what can and cannot be included. Some assumptions are made and warning labels and caveats are used where the information is lacking or incomplete – a series of red / yellow flags appears.

This approach could also be used in the National Climate Assessment.

Overall recommendations from this session include
- We should try to take advantage of existing distributed data and metadata systems, including those not explicitly presented at this workshop. However, many of the data sources that we need (especially regarding impacts and vulnerability) may be outside the government, and IQA requirements call for the validation of their accuracy.
- These existing sources also need to meet their primary objectives as determined by the agency or other funder. The Assessment/ USGCRP could help provide a consistent framework for connecting the relevant data systems.
- Inter-operability, keeping the data close to the experts, procedures for maintaining data, and other long-term support (funding) for the researchers are all essential and require significant attention to detail in designing the knowledge system.

Value of Information – Establishing Priorities for the NCA

This presentation addressed the issue of prioritization of data sources. Since there is an overwhelming amount of data available today, and the volume will only increase over time, it is important to assess which kinds of information are truly valuable and how those kinds of information can best be deployed. Ways of thinking about prioritizing the data management activities for the Assessment included
- If we cannot take action in response to information, it has limited value to decision makers.
- Information costs money. How to do we decide what is most important?
- The value of information may not justify the cost of its acquisition.
- "Reducing uncertainty" should not be a metric. Sometimes collecting information will increase uncertainty; we should not create false expectations.
- Some challenges:
  - Information is a public good when it is generated by the government – things that are free are often not valued, even if they are actually very valuable. And those who

derive value may well not be those paying the bills for the data or the data management. This makes it hard to protect government data programs from cancellation.

- o Where public goods are used for private benefit, equity issues can arise.
- o How much information is enough? What quality and resolution (spatial, spectral, and temporal) should we focus on? The answers to these questions may vary dramatically depending on which research questions we are trying to answer.

- Some methods for making decisions between the alternatives:
  - o Price and cost based derivation
  - o Probabilistic approaches
  - o Regulatory cost effectiveness
  - o Econometric modeling and estimation
  - o Simulation modeling and estimation
  - o Increasing the value - need to demonstrate willingness to pay for information
  - o Use the valuation exercise to think through data collection and assembly in a structured way and prioritize activities.

Input from Breakout Sessions

Data Needs and Challenges

Unconventional data sets, such as those coming from insurance companies and health departments, present a significant problem. This information is extremely useful, but will require special procedures. We will need to solicit input on what types of data we should be collecting and monitoring, and then figure out how to minimize risk in using non-traditional sources (*e.g.*, state, non-governmental organizations (NGOs), and academic data). There are many other databases out there that are compilations of reputable, federal datasets. Any data that we use will have to pass the "transparency test". However, this is very difficult with socioeconomic data, traditional knowledge, and impact data. How will we ensure quality? In 2007 the *Evaluating Global Change Assessments* report from the National Research Council recommended formation of a *nested matrix* where

a basic data framework would be developed for the whole country, with opportunities to perform more in-depth analysis/assessment in regions or sectors within this broader assessment framework. This would overcome the problem of needing very large quantities of data to do local analyses across the whole country in the near term.

Overall input from this session included

- Physical data challenges are trivial compared to working with anthropic data and socioeconomic data.
- It is critical to understand how people are using the data, what data they really need and at what resolution.
- We may need an iterative process to identify and gather the data that we need over time.
- Communities of researchers are using major data archives such as the model output from AR4 far more than was originally anticipated. These community efforts should be encouraged.
- We will need to have clear documentation about the data; how it was produced, strengths and limitations, why it was generated and why did the principal investigator think it was relevant and useful for that particular part of the assessment?

Criteria will be needed for including new data and information as official Assessment data, (*e.g.*, categories of quality or dataset maturity, scale issues – temporal and spatial). One priority is documenting changes that have occurred since the last time an Assessment report was written (*e.g.*, impacts). We need a way to document what is changing in a standard and consistent way through the use of indicators.

Organizing the data is difficult – should it be done at a national level, *e.g.*, the USA National Phenology Network; at the sector level, *e.g.* through professional disciplines such as health; through online journals that provide open access? The new Assessment process includes a two-pronged approach, a dynamic approach that is providing up-to-date information from the Assessment process in near-real time in a web-based format, and a static approach that is the report or reports, which act as a slice through the process where we will need to preserve the data that were used to create it. This involves all sorts of qualitative as well as quantitative information.

We should try to take advantage of distributed data systems:

- There has been an inter-agency group (the Climate Change Adaptation Working Group (CCAWG) which includes the U.S. Army Corps of Engineers, the Bureau of Reclamation, and NOAA among others) working on indicators for water; they have identified sources of information and where there are gaps in the databases.
- There are many existing databases and data systems that we could draw upon. Experts in a topic can quickly identify many of these sources for a particular sector.
- Historical climate information has always come from NOAA's National Climatic Data Center (NCDC), which has played a major role in both previous National Assessments. Remote sensing could bring NASA into the picture more while projections of change could come from Lawrence Livermore National Laboratory and NOAA. What can be done to organize this in the early stages to pay off in the long term?
- We will need to solicit input on what types of data we should be collecting and monitoring. For example, economic, ecological, and other data types to document global change and underlying vulnerability. Many of these sources may be outside the government.
- There are specific needs for economic, social science, and impacts data.
- Proxy sources may be valuable. For example, trade associations, such as local wine growers, may have data that integrate past climate conditions relative to agricultural crops.
- Department of Energy (DOE) and others have done some work on impacts data (Vallario at DOE and de Sherbinin at SEDAC, Socioeconomic Data and Applications Center)

Indicators

When looking at extremes (e.g., drought), there has not been particular consistency in the use of indicators (should we use number of days of drought? intensity of drought? geographical extent of drought?). There also has not been consistency on what the period of human influence is. There is often a significant difference between basic science indicators and impact indicators. Sectors can help identify what the indicators are that stakeholders

care about, as well as sources of information for these topics of conversation.

Some indicator ideas include

- There should be a principle of parsimony and of realism. We should avoid using too many indicators. It is a lot of work to measure all of these. One criterion for indicators is that they need to be measurable.
- Conduct a review of indicators. Find out what indicators are already in use and already being collected.
- There is a need to dissociate climate impacts from local environmental issues (e.g., land use change that is not related to climate change).
- There is potential for use of data.gov as a way of visualizing data and indicators.
- Some indicators are easily accessible and widely understood by general populations (e.g., number of frost free days in a year). Other indicators are more analytic and more difficult for non-scientists to understand. There needs to be balance between complex analytical indicators and indicators that are easily accessible by the public.
- The Assessment needs to group indicators such as global vs. local, within sectors to document the scale of the impacts and observations that are represented by each indicator.

Criteria for choosing indicators can include cost to produce, presence of peer-reviewed literature showing clear relationships with climate, flexible, evolutionary, integrated, useful/relevant, parsimonious, tiers (or nested) indicators, common sense indicators that are easily explained.

Criteria, Special Considerations, and Conditions Related to Different Types of Source Materials

There are significant opportunities as well as concerns related to documenting vulnerability to climate impacts as well as evaluating adaptation and mitigation strategies. There is very little literature on these topics; thus the Assessment is likely to generate important information that needs careful documentation. Methodologies used in social sciences provide more than just anecdotal evidence – we need to engage with social science communities to determine appropriate standards from the disciplines of geography, anthropology, sociology, *etc*. We also need to engage with other professional communities (*e.g.*, mayors, agricultural interests) that generate and use data in operational settings – and many of them do have trade journals and review standards. Can we use anecdotal evidence to help us set place-markers for further investigation or as a component of assessment that brings forward "other voices"? Will it help determine who is disproportionately affected by climate change? Precedent / guidance on this could be provided by the Environmental Protection Agency (EPA).

This session included a discussion of general criteria for including new data and information

- Data of questionable quality, mistakes, or misuse of data could threaten the whole Assessment process. On the other hand, we should not fear new data sources which may be extremely valuable. We will have the authors themselves, the Federal Advisory Committee, the NRC, and other review processes to ensure data quality.
- We will need a clear and relatively simple way to communicate about data quality and scientific uncertainty; data identifiers and metadata are essential.
- There are specific needs for economic, social science, and impacts data.
- Scale concerns may be addressed by the nested matrix concept (more detail in some regions nested in a broader information base).
- It may be necessary to give more weight to the data that has been collected for longer periods in some cases, because trends may be better documented and continuity of data generally leads to higher credibility
- Data from sources with rigorous and regular QA/QC (quality assurance/quality control) will need to be given the greatest weight.
- Is new data providing new insights? Does it provide something that historical data does not? Or does it further corroborate existing data streams for added statistical power?
- Local scale data could be important, as long as it is relevant to climate issues.
- What about insurance industry data that we have access to, and could be very useful, but cannot be released because it is proprietary? Is there any way to summarize this information for Assessment purposes?

- Some human-subjects data cannot be released due to legal requirements to protect the subjects; however, data can be repackaged in ways that protect the identity of the subjects, and then they become usable and publishable. Similar issues often apply to threatened and endangered species data.
- Some data may come from commercial sources (such as private-sourced airborne and satellite-based imagery) as well as from missions of other nations. These data may be highly useful in an assessment, but may not have redistribution rights, which can present unacceptable limitations on transparency.

Thoughts about Peer Review

How should the Assessment cite peer-reviewed papers? Do we need to have a database of the data used to create these papers or is a reference list enough? If things have already been published in peer-reviewed journals, the NCA should not have the obligation to obtain the original data and make it available.

Professional societies, *etc.* may be able to help with publishing data so that it meets the peer review criteria. There are several on-line journals in place that allow publishing data that can then be cited. We should encourage this and help to establish new tools to help Assessment researchers get their work in press quickly. This would be cost effective.

Economic costs of climate change are very hard to document. The ongoing costs of climate risk management (snow removal, *etc.*), and also costs of acting (adaptation/mitigation costs) are very subjective. Numbers can be estimated, but their accuracy is questionable. When we are making a summary or giving policy advice based on data, we need to make sure that the data is very strong.

We need to find a balance between getting in all of the useful information and having a functional and manageable peer-review process. In general we are taking a high-level national approach, but in many cases local data will be useful.

To what extent does everything have to be amenable to the peer-review process? What about contrarian views that claim they are stymied by the peer-review process?

We have to be careful not to just use non-peer-reviewed perspectives because it supports the narrative (*e.g.*, why include the indigenous perspective from Alaska if it does not have broad applicability across the nation? Are we opening ourselves to criticism by focusing on those who are most vulnerable?). However, if we include this information and make a different "category" for it, that may not be appropriate either.

Case studies *vs*. anecdotal evidence: Should we involve anthropologists, sociologists, *etc*. to help vet socioeconomic and cultural data? They can say whether it was rigorously collected and meets professional standards. Non-traditional professional/technical sources of data, including societies like the Council of Mayors, *etc*. have a lot of statistics from their areas, and may have long-term data on things that could be relevant (*e.g.*, budget spent each year on plowing, incidents reported in heat waves).

"Grades" of peer review: What if the information/data are published in a trade magazine? Even if there is peer review in the trade magazine, is it as good as articles published in academic journals? This will depend on what the data are being used for. The NCA should not base a major conclusion solely on something that was peer reviewed for a trade magazine, but it could support a minor conclusion or recommend that it be separately peer reviewed for Assessment purposes.

Scale Concerns and the "Nested Matrix" Concept

- At what scale do we need data? What if we can only get certain types of information in certain regions or sectors? Having illustrative examples may be useful, and we need to allow for regional differences in what data is actually relevant in certain places. The NRC has recommended a *nested matrix* approach for concentrating resources and to have more depth in our analyses.
- Throughout our creation of the indicators and the "national matrix of indicators", we need to keep in mind that the climate itself and climate change varies by regions.

Data Quality and Characterization of Uncertainty

- We will need a clear and relatively simple way to communicate about data quality

and scientific uncertainty (*e.g.*, a "progress bar" or "state of knowledge" about our understanding of a topic and how it has changed over time) - for the public, for policymakers, for other decision makers, *etc*., who all have different needs and levels of understanding.
- Some sources are reliable partners. Others are less so or are unconventional (*e.g.*, state and local data, industry data). In either case, the Assessment will need to lay out criteria for long-term curation of data. This will be a data quality challenge especially for the ongoing Assessment process. The path is clearer for the report itself.
- How do we make sure that our processes are robust enough that our use of data that are not published does not threaten the Assessment process?
  - o Inappropriate use of peer-reviewed scientific information is always a risk, too. Documenting the intended use of the data would help limit misuse.
- Transparency, provenance, and availability of "litmus tests" for non-federal data sets are criteria to be considered. There will be challenges for proprietary private sector information, for example.
- Does the National Climate Assessment have a responsibility to describe data quality standards and "best practices" for various methodologies, and share these with non-governmental information providers? An example is the NOAA IOS (Integrated Observing System) Team which has an accreditation process and brings in private sector and community data.
- Need a clear and simple way to talk about data quality and uncertainty. This might need to be tailored to different audiences.
- Data identifiers are essential for accessing the metadata that will explain who created the data, how it has been stored, and why it was created.

2013 Report and Web Deployment

The Assessment report introduction will need to describe the continuing process as well as the product-as-a slice-through-a-process theme. Each appendix will also need to document the process and the workshops that generated the ideas that were considered, and supporting documents will be available on or through the Assessment website.

- Need multiple portals / ways into the data depending on audience
  - Building the Assessment (scientists, authors, *etc.* need web-based places to share their information)
  - Databases and links to data on other sites
  - Atlas (well developed narratives and documents that can be referenced to a map)
  - Decision makers (a site focused on adaptation and mitigation decisions)
  - Trainers and educators (a site for sophisticated users who need a different level of information)

The written report due in 2013 should be a foundation based on science. Beyond the initial report there can be additional products that can be interpreted and reformatted for public consumption. The actual report should have lots of citations and be heavily peer reviewed.

Not everything that we will include in the Assessment will make it into the peer-reviewed literature. Therefore, we need to set our own criteria and figure out our own review process.

Do we need a peer review team? If so, should that team be part of the FAC or separate? To make sure we have transparency, maintain the level of documentation that we would like, and to facilitate the use of data generated by stakeholders, we will need a strong and defensible review process. This could be more rigorous than regular peer review. It is a structural question - how to ensure that all of our different teams meet our standards for knowledge management.

Some individual input related to peer-review issues include

- The Assessment could have an ombudsperson to resolve conflicts concerning perceptions, actual conflicts, *etc.* This person could serve on the FAC.
- To meet IQA standards, does everything need to be documented and peer reviewed in the same way? Or can some information be flagged as "preliminary" or "advisory only" while other data are considered "reliable" or "robust"?
- Online peer-reviewed data for IPCC has been discussed. Supporting societies such as the American Geophysical Union (AGU) can

publish either data or articles – it benefits them, too. The NCA's challenge is lack of staff and resources – we cannot pursue publication of all the data we need ourselves. If there were a community of people who could help make this happen, it would be a reasonable component of our data management approach.

- Even traditional peer review can be flawed. We do not want to hide behind the peer-review process. The fundamental focus should be on verifying/evaluating the facts and adaptive learning over time.
- The NRC will also review the reports, so if we have a peer review committee **and** the NRC for review we will be more secure that the information we have used is reliable.
- Chain of custody of data (who collected it, who analyzed it, *etc.*) is an important documentation issue. Labeling properly can provide guidance on what to expect of the data.
- How do we deal with a topic where there is only one study available – can a conclusion be based on one study? How do we create an executive summary that captures the discussion from all angles that are provided in the report? It can be quite confusing when trying to present data or findings from various levels of review – we need to have some sort of minimum standard and a method for making sure that the report is "readable."

Peer Review Strategy for the Ongoing Assessment Process

- An audit of whether the review process is adequate to the task can be conducted on an ongoing basis; there is literature on how to do this.
- Have peer review, community review, stakeholder review, and agency review of the Assessment process itself - especially involving people who may have strongly held views. It may help to legitimize whatever comes out of the report.
- Decide how to capture data and other background papers that were used to produce an official Assessment Report.
- Do not let the authors choose their reviewers, have a conversation about this up front with the authors.
- Remember that with the proliferation of journals, everything is publishable these days – need some other people who have broad experience in peer review to act as moderators.

- Peer review should be conducted at the start (is the process adequate to achieve the desired results?) and at the finish (was the process followed and what worked and what did not work?). We plan to use these process workshops and reports from them to develop input to the process and the FAC will provide oversight of this. Let people know what this process is up front. Methodology will be clearly defined / explained within appendices to the report.
- Follow the Bromley Principles (early USGCRP data management principles when Dr. Allan Bromley was at OSTP – can be found at http://www.gcrio.org/USGCRP/DataPolicy.html). We need to document our standards.
- Make sure that whatever the data policy is, it is prominently posted / available and that "chain of custody" is well documented. Make sure author teams are amenable to the forms of peer review that are required, since their work will be subject to that standard.
- Maintain rigor in underlying research and how research is applied to reach conclusions (two main categories addressed in review process of reports). With the second category, care must be taken regarding the degree of certainty ascribed to any one particular study.

Public Comment

- Public review: how to make sure that we respond to the most important comments and do not get bogged down by the sheer number of comments.
- There was a good process for IQA/public comment on the last Synthesis and Assessment Products (SAPs) by the Climate Change Science Program (CCSP). Could evaluate the agency public comment processes for ideas.
- What if we get 90,000 comments? How would we handle this – we cannot really respond to each individually. There is a good chance that we will get a lot of comments. Even the 4,000 comments that we received on the 2009 report was a lot of work.
- There are software packages for analyzing the comments that come in – this enables us to distinguish between unique comments and form letters. Also, there is precedent for hiring contractors to summarize the comments into categories to manage the workload.
- We are currently using the same comment architecture at USGCRP that IPCC used, but it does not have searching capabilities.

- There should be guidelines about what constitutes an adequate response to both public comments and peer review comments. We can design these guidelines with an estimate of the workload in mind.
- What do people think of flagging the data according to our level of confidence/data quality levels… seems like a judgment call, should we do this?
- Responding to public comment can require an inordinate amount of time of those with the scientific expertise. This needs to be carefully managed.
- What about input into what we assess – sectoral experts need to help define what is assessed. We are already soliciting some of this input via the Federal Register Notice (comments due October 8, 2010).
- Increasing calls for transparency – but this also comes with increased administrative overhead. Is there a "trigger" that would tell us we need to review the process because there is too much administrative overhead?
- Could the communications strategy be used to referee some of the interaction with different publics?

Archive and Access for Assessments: Documenting Data and Source Materials

We will need to document our procedures for building and managing the data system (*e.g.*, chain of custody, quality control, approach to shared databases, archiving, and security)

- There are different documentation needs for the report and for the ongoing process. How will these parallel processes be managed?

Putting the report on the web is a bare minimum; we need multiple work products and processes on the web:

- "About the Assessment" on globalchange.gov focuses on the Assessment process.
- Workgroup sites for collaboration: could be PBworks or Oracle framework – ORNL has used Webex. Cybersecurity concerns need to be addressed for access from non-government partners.
- Delivery of Assessment products (could Climate.gov be a host for some of these?): describe Assessment products that other people can build off of and create front-end services. Web services requirements (*e.g.*, Open Archive Initiatives-Protocol for

Metadata Harvesting (OAI-PMH)) should be set out by the Assessment, and enable other people to build tools from this framework/infrastructure.

- The Assessment Indicators and the Assessment Data
    - Lay out some basic criteria and processes for the inclusion of information (may be iterative); the durability of data will be a challenge.
    - There are many examples we can learn from: federal agencies and interagency programs, citizen science.
    - For monitoring systems: Are there peer review standards already in place?
    - Underlying data and transformations / analyses must be transparent or accessible to the public (implications for using proprietary data? implications for using social / human subjects data?).
    - Is there a minimum/desired period of record for data sets? This may vary based on the issue.
- Interpretations of the data: What new insights do these data provide? Do they corroborate existing data sets or add richness (e.g., greater temporal or spatial resolution)?
    - How the data are used is critical – threshold of validation (i.e., does a key conclusion rest on the data? If so, the review standard should be very strict).

Quality vs. Quantity and Relevance vs. Detail

Does everything done need to be on the web and be accessible to everyone? If having all the data makes it confusing to the public, we may want to identify what is most appropriate to provide. Transparency is not the same as posting everything – there is a balance between helping people understand the process and giving them access to the people who did the work vs. transparency.

Metadata

The vocabulary of metadata needs to be standardized - metadata is the foundation for accountability. Metadata is all that information about the data short of the actual value (e.g., column headings; date, time, and location of collection; handling information).

Multiple metadata standards are already out there – the key is to have the information about where the data came from in a standard format; it can then be translated amongst these standards if needed.

- Two major purposes (1) to support why the data were generated, how it should be used, how the data was culled, etc. (2) to give people enough information to decide if they should use that data for their purposes
- Supporting reasons for using particular data – any standards followed, collection information, what was discarded and why / what data were kept and why, processing of data, accuracy and precision of observations (also, of locations / collection points), minimums/maximums allowed, other information / categories (state names, time periods, etc.)
- Can document who the data collectors were, whether they meet the requirements of the IQA, statements about appropriate use (or inappropriate use), disclaimers
- Metadata are used to link narratives back to the actual data/link graphs back to source data; using metadata is both feasible and desirable. For example, graphs/narratives on climate.gov need to include links to source data, and links to related sources and information.
- Metadata that integrate climate data with geographic data is important, though the climate- science community does not necessarily value of exact location of data. E.g., Agencies need to make sure when they collect location of data, they state how that location was determined (e.g., GPS).
- The use of derived products is accelerating. For example, there is now a website in the Southeast that predicts and keeps track of frost days, which is usable by peach farmers.
- Metadata will be made available, but of course there is always the concern that people will not look at it, and will still use the data inappropriately.

In FGDC (Federal Geographic Data Committee) metadata format there is a requirement to enter "use constraint". Also, within National Ocean Service, they are required to put in disclaimers and usage constraints. We should try to highlight these key use constraints.

At regional climate centers, users typically do not want to see metadata. When asked, most say "no"; they do not want metadata. Users will do what they want with our information, but we still need to use metadata to fully document everything. Even changes in instrumentation over time may not have been documented. We need to be the best broker of information possible, but then realize people are not going to always know how to use the data and understand its caveats.

- Practical aspects – make metadata a part of normal work flow. It should not be written by just one person (collectors, processors, *etc*.). Write down processes for people collecting and processing data (including practical considerations like where to submit, who to call with problems, *etc*.).
- Information systems – what fields should be searchable (these will have to be in the metadata). For distributed systems, remember that different places use different fields and interpretations; we need to make sure everyone does things the same. Be sure to define terms and acronyms. Do not forget to think about the programming required for a data management system.
- In the metadata, there needs to be a general description of people who touched the data. Note whether they were well-trained people in a certain field.

Metadata for Data Discovery – Search Functions and Implementation Ideas

Metadata describe the data set: key words, geographical area of data, *etc*. This is typically the information someone looks at to see if it is of interest to them. The title is the first thing presented to someone; it should be descriptive (*e.g.*, topic, date, and location).

- Should we put a tab on the website that says "for scientists" (find more technical information here)?

- The web-based portion should probably go beyond the report structure and be more interactive (*i.e.*, not just a PDF of the report).
- Is there value to engaging the public on the contents of the document (*e.g.*, public forums)? To what extent should this be building and supporting a community?
- Should we set up a living metadata catalogue?
- A decision matrix for what data to accept? How do we say "no" to some people?
- Start with obvious datasets, and add from there according to the decision matrix.
- Need to make sure to define authoritative data sets that need to be included.
- References will be needed on the website; not all references are openly available (need journal access). Is there a way to have the PDFs of all of the references? This is a big challenge, since a lot of journals require subscriptions to see them. We should start working on this right away.
- A lot of systems are moving towards open source to encourage interoperability.
- Should we use strategic partners to develop the website (*e.g.*, Google, NCDC, U.S. Global Earth Observations System of Systems (USGEOSS), or professional societies)?
- We need to make sure that years from now people will be able to reproduce what is in the report. They need the data. Data that were used to generate the report needs to be preserved even though the ongoing process will continue. There needs to be a cutoff date for what data will be accepted in a particular report.
- Are we required to capture these data and keep it forever?
- Identify key strategic partners inside and outside of government
    o Content and infrastructure providers (into the Assessment)
    o Translators and champions (out of the Assessment)
    o Co-producers (with the Assessment)
- Prioritize features for the website (what we provide now, what we hope to provide later, what we hope partners can provide now or in the future).

The Assessment is working closely with NOAA Climate Services, regional centers at NOAA and DOI (Department of Interior), and will keep in mind USDA's (U.S. Department of Agriculture) capabilities and connect to their data experts. Integration between the agencies is very important, and they are already working toward this end through sharing data, using open source, *etc*. NBII (National Biological Information Infrastructure) at USGS has metadata capabilities. DataOne has some capabilities we should leverage also.

We should use infrastructure that has already been developed when possible. We also should leverage scientific teams to contribute to the public interface.

Who is the Audience/User?
- Need to keep in mind that we are not creating climate services, though we are intending to support them. We should aim to serve sophisticated users who can then translate to users such as "on the ground" managers. The general public is not our primary user.
- Develop a user model for public use of the site. How much can we spend on each component (how much to spend on which aspects of the site)? Drupal, a content management system, can be a useful tool.
- User profile analysis and requirements: who do we mean by policy makers, and what do they want from the site? Part of building relationships with these users/audiences/ collaborators. Since there will be new users, there is an outreach aspect to this work. This process guides the content for the site. NCDC is doing this for climate. gov and will be presenting on this topic at the November USGCRP communications workshop.
- Leverage Regional Integrated Science and Assessments (RISA) programs, Climate Science Centers (CSC), Landscape Conservation Cooperatives (LCC), and other regional centers that can access users outside the government. RISAs may also be able to create the tools for users, based on what the users tell the RISAs they need (and not the other way around).
- This process can help us avoid trying to be everything to everyone, or serving the general public.

Core Team Requirements

A preliminary conversation about the possible roles and responsibilities of people who were present at the workshop included the following possible contributions

- The DOI data architecture team
- Peter Murdoch's Climate Effects Network (CEN) at USGS
- Metadata experts (e.g., Anne Ball, NOAA; Ted Habermann, NOAA)
- Web interface (John Keck, NOAA/NCDC)
- Federal employees who know what is already going on with data management within the federal government (data systems). What are the key assets, and who we should connect to in these agencies? (DOE's science office) (Glenn Rutledge, NOAA; Bruce Wilson, ORNL; Dean Williams, PCMDI; Bob Chen, CIESIN; Chris Lynnes , NASA; someone from HUD)
- Coordinator function (someone in the USGCRP office)(probably a UCAR hire because we need a long-term system that is consistently managed)
- Distributed expertise – including people who are embedded in regions and sectors. This will leverage off of the Assessment regional and sectoral leads and their contacts (National Institute for Food and Agriculture at USDA, DOI Climate Science Centers and Landscape Conservation Centers, new regional directors for the Climate Service at NOAA, etc)
- Digital librarian who understands how content pieces and types of data fit together (Bruce Wilson, ORNL)
- Managing editor to set tone and style (NOAA/NCDC)
- Cybersecurity and federal requirements (someone who will not just say NO but facilitate good outcomes – OIRA/OMB?)
- Usability and assessment of interface and tools – to scientifically assess whether or not we are meeting user needs (Sea Grant, UTK – Bruce Wilson knows a little bit, RISAs)
- Scientific content and communications experts (Ned Gardiner, NOAA/NCDC)
- GIS expertise (georectification expertise), georeferencing for our complicated geographical information (watersheds, ecosystems) (Ned Gardiner, NOAA/NCDC; USGS; NGDC)

Accessing Data and Source Materials on the Web

Who will manage the real-time data being displayed on the website – with someone doing peer review or data quality checks on an ongoing basis? There is a tradeoff between accuracy, relevance and timeliness *vs.* workload tradeoffs. We would like to bring people new science as it comes, but how can we actually meet the challenges associated with doing this?

- Assessment will provide authoritative data.
- Centralized infrastructure may be something that we can leverage using open standards, with a distributed system (comparable to IPCC and climate.gov's architecture).
- USGCRP strategic planning should help clarify the roles of the Assessment, climate services, and adaptation activities. This may help us understand the boundaries between Assessment ("wholesale") and third parties providing translation?
- Should we just put a static report on the web or move toward having a dynamic web site?
  - o No one seems to think that a static report on the web is a good idea, but there are concerns about resources for maintaining a dynamic site
- Who should we be reaching?
  - o Science / author community (those building the assessment) - ingesting the information
  - o Broader audience (those consuming the assessment) - dispersing the information (range of stakeholders / users each with own needs – education, communication, decision making)
- Build an architecture that serves those contributing and that could be built out for larger science community
- Customizable content – publicly accessible portal(s), plan for engaging communities to use this portal (what tools are most useful and understandable – maps, graphs, *etc.*)
- Building a virtual community to discuss findings

Links to NCA Communications and Engagement Strategies

We are trying to serve various populations: decision makers with specific needs in specific locations, national climate service, levels of government, and sectors. We will need to identify our priority audiences as our foundation, and then add on more users/audiences after we have tried to serve our priority audience.

Who is responsible for the marketing of the NCA? In the past, it has been an afterthought. This time, we would like to embed the communications strategy into the process. Communications are critical to success due to the distributed process – within and across agencies, and in connecting to Assessment partners. This is not just about communicating results.

Role of the Web in Communication Strategy

- The web is one way of communicating, not an end in and of itself.
- Figure out ways to get the Assessment represented on climate.gov and other sites
- Provide content to the education part of climate.gov (noting that education per se cannot be one of our highest priorities in the short term due to resource limitations.
- Make connections through social media - consider development of mobile applications – the web is not the only way of providing access. *E.g.,* an application that points to "what's new" on the Assessment – a pointing tool.
- Publish some standards for application developers. (Climate.gov will have a mobile application)
  - o The federal government has to maintain standards of open information access to all. It is not wise to create a mobile application for a specific device. At least if it is just on the web, people have free access at public libraries, *etc.*
- Need a targeted media strategy when the report comes out. *E.g.,* The Union of Concerned Scientists broke down the last report into small summaries that were used by local media and others. We should involve stakeholders in the release strategy.

- With the last report, the White House helped digest the report – which was great for a one-day national news splash, but then subsequently interest diminished. There was also a regional release strategy that was canceled at the last minute; it could be of use in this next report release.
- We need both short- and long-term measures of success for dissemination of content from producers to users of information
  - Short-term: news / blog stories, inquiries to the NCA Office, *etc.*
  - Long-term: use in planning and decision making processes; co-production of information with stakeholders in regions and sectors, *etc.*
- The Web portal could have two segments – a data segment (references, *etc.*) and a more publicly accessible segment for decision support, *etc.*

Appendix A: Agenda

Monday, Sept 20

12:30pm-1:15pm: **Arrival** (Please leave sufficient time to get through security prior to meeting)
1:30pm: **Welcome and Overview of Assessment** – Katharine Jacobs, OSTP

1:50pm: **Overview of the knowledge management challenge, desired outcomes for the workshop** – Anne Waple, NOAA

2:15pm: **PANEL - General knowledge management challenges/solutions from previous assessments.**
- 2:15-2:35pm - U.S. Assessments - Tony Janetos, UMD
- 2:35-2:55pm – IPCC, including the IAC recent review – Greg Symmes, NAS
- 2:55-3:15pm – Annual State of the Climate reports- Deke Arndt, NOAA
- 3:15-3:45pm - Plenary questions to panel – moderator, Kathy Jacobs, OSTP

3:45pm: **Break**

4:00pm: **Information Quality Act and Highly Influential Scientific Assessments** – Nancy Beck, OMB

4:20pm: **Plenary discussion** – moderator, Kathy Jacobs, OSTP

5:00pm: **Summary and charge for next day and a half, general overview of breakout plans** – Anne Waple, NOAA

Tuesday, Sept 21

8:00am-9:00am: **Bagels, pastries and coffee in meeting room**
                          (Please leave sufficient time to get through security prior to meeting)

9:00am: **Peer review of assessment materials and data in plenary** – Kathy Jacobs, OSTP
                          Plenary discussion will include:
                          1) Peer review strategy for national assessment report (every 4 years, delivered to Congress)
                          2) Peer review strategy for ongoing assessment process (rolling release of sub-components)
                          3) Responsibility for peer review – Academy, blue ribbon panel *etc*.
                          4) Handling public input and how we respond to public input
10:05am: **Instructions to breakouts** – Anne Waple, NOAA

10:15am: **Break**

10:30am: **Breakout A (2 groups), each discussing input related to:**
1) Data, information sources needed to support the 'National Matrix' and evaluating impacts on regions and sectors (including common indicators)
2) General criteria for including new data and information (categories of quality or dataset maturity, scale issues -temporal and spatial, etc)
3) Functionality of data system (*e.g.,* chain of custody, shared databases, archiving and security)

12:00pm: **Lunch -**USGS Cafeteria

1:00pm: **Keynote: The Value of Information - Molly Macauley, RFF**
Including plenary questions/discussion- moderator Kathy Jacobs

1:30pm: **Breakouts report back to group** (rapporteurs report out, moderator – Anne Waple)
• 10 minutes for each rapporteur

1:50pm: **Plenary Discussion**

2:30pm: **Break**

2:45pm: **Lessons learned in NIDIS** (and plenary Q&A) **– Roger Pulwarty, NOAA**

3:15pm: **Overview on access/delivery challenges for the Assessment–** Kathy Jacobs, OSTP

3:30pm: **PANEL: Providing access to the National Climate Assessment and its sources**
• 3:35pm –The role of Climate.gov - David Herring, NOAA
• 3:50pm – IPCC data/information access – Dean Williams, PCMDI
• 4:05pm –Metadata in a data/information service - Anne Ball, Coastal Services Center, NOAA
• 4:20pm - Plenary discussion - moderator Ned Gardiner

5:00pm: **Adjourn**

Wednesday, Sept 22

8:00am-9:00am: **Bagels, pastries and coffee in meeting room**
(Please leave sufficient time to get through security prior to meeting)

9:00am: **World Bank Climate Portal – an overview** - Ian Noble, World Bank

9:30am: **Summary and introduction of breakout topic –** Anne Waple, NOAA

9:45am: **Breakout B (2 groups)- Information Base for the National Assessment**
1) General requirements and critical features for (web-based) information base (*e.g.* spectrum from static summary of report to dynamic, semantic searchability and map interface)
2) Assessment information to service and products – how to integrate
3) Role of the web in communication strategy
4) Building from existing federal and non-federal resources

11:00am: **Break**

11:15am: **Brief report back from breakouts (rapporteurs) and plenary discussion –** moderator Anne Waple, NOAA

12:00pm: **Summary of workshop and next steps –** Kathy Jacobs, OSTP

12:30pm: **Adjourn**

Appendix B: Participant List

Derek Arndt
NOAA
derek.arndt@noaa.gov

John Balbus
National Institutes of Health
john.balbus@nih.gov

Anne Ball
NOAA
anne.ball@noaa.gov

Nancy Beck
Office of Management & Budget
nancy_beck@omb.eop.gov

Nancy Beller-Simms
NOAA
nancy.beller-simms@noaa.gov

Ken Belongia
US Geological Survey
kbelongia@usgs.gov

David Blodgett
Wisconsin Water Science Center
dblodgett@usgs.gov

Mary Boatman
Office of Science & Technology Policy
mary_c._boatman@ostp.eop.gov

Jeffrey Budai
NOAA
jeff.budai@noaa.gov

Emily Cloyd
USGCRP
ecloyd@usgcrp.gov

Bernetta Crutcher
Dept. of Transportation
bernetta.crutcher@dot.gov

Alex de Sherbinin
Columbia University
adesherbinin@ciesin.columbia.edu

Ned Gardiner
NOAA
ned.gardiner@noaa.gov

William Goran
Army DOD
william.d.goran@usace.army.mil

Linda Gundersen
US Geological Survey
lgundersen@usgs.gov

David Halpern
NASA
david.halpern@nasa.gov

David Herring
NOAA
david.herring@noaa.gov

Katharine Jacobs
Office of Science & Technology Policy
kjacobs@ostp.eop.gov

Jenna Jadin
USGCRP
jjadin@usgcrp.gov

Anthony Janetos
University of Maryland
anthony.janetos@pnl.gov

Lesley Jantarasami
EPA
jantarasami.lesley@epa.gov

John Keck
NOAA
john.keck@noaa.gov

Stuart Levenbach
Office of Management & Budget
stuart_levenbach@omb.eop.gov

Neal Lott
NOAA
neal.lott@noaa.gov

Amy Luers
Google
amyluers@google.com

Christopher Lynnes
NASA
christopher.s.lynnes@nasa.gov

Molly Macauley
Resources for the Future
macauley@rff.org

Renee McPherson
University of Oklahoma
Southern Climate Impacts Planning Program
renee@ou.edu

Martha Maiden
NASA
martha.e.maiden@nasa.gov

Cheryl Morris
USGS
cmorris@usgs.gov

Jarvis Moyers
NSF
jmoyers@nsf.gov

Ian Noble
World Bank
inoble@worldbank.org

Jim O'Brien
Florida State University
Southeast Climate Consortium
jim.obrien@coaps.fsu.edu

Sheila O'Brien
USGCRP
sobrien@usgcrp.gov

Carolyn Olson
USDA
carolyn.olson@wdc.usda.gov

Jennifer Parker
CDC
jdparker@cdc.gov

Adam Parris
NOAA
adam.parris@noaa.gov

Roger Pulwarty
NOAA
roger.pulwarty@noaa.gov

Mohan Ramamurthy
UCAR
mohan@ucar.edu

Carolyn Reid
US Geological Survey
clreid@usgs.gov

Chris Rewerts
US Army
chris.rewerts@us.army.mil

Kevin Robbins
Louisiana State University
Southern Regional Climate Center
krobbins@srcc.lsu.edu

James Rolfes
Department of Interior
james_rolfes@ios.doi.gov

Glenn Rutledge
NOAA
glenn.rutledge@noaa.gov

Paul Schramm
USGCRP
pschramm@usgcrp.gov
Aaron Smith
USGCRP
asmith@usgcrp.gov

William Solecki
City University of New York
wsolecki@hunter.cuny.edu

Brooke Stewart
NOAA
brooke.stewart@noaa.gov

Alan Strasser
Dept. of Transportation
alan.strasser@dot.gov

Greg Symmes
National Academy of Sciences
gsymmes@nas.edu

Michael Tanner
NOAA
michael.tanner@noaa.gov

Janet Tilley
USGS
jtilley@usgs.gov

Monica Tomosy
US Forest Service
mstomosy@fs.fed.us

Lucia Tsaoussi
NASA
lucia.s.tsaoussi@nasa.gov

John Walsh
University of Alaska at Fairbanks
jwalsh@iarc.uaf.edu

Anne Waple
NOAA
anne.waple@noaa.gov

Dean Williams
Lawrence Livermore National Laboratory
williams13@llnl.gov

Bruce Wilson
Oak Ridge National Laboratory
wilsonbe@ornl.gov